YOUR KNOWLEDGE HAS VALUE

AF144004

- We will publish your bachelor's and master's thesis, essays and papers

- Your own eBook and book - sold worldwide in all relevant shops

- Earn money with each sale

Upload your text at www.GRIN.com
and publish for free

K.R. Sasitharan

Implementation of P&O MPPT based Zeta Converter fed

Three-phase Induction Motor

GRIN Publishing

Imprint:

Copyright © 2015 GRIN Verlag, Open Publishing GmbH
Print and binding: Books on Demand GmbH, Norderstedt Germany
ISBN: 978-3-656-97349-2

This book at GRIN:

http://www.grin.com/en/e-book/295167/implementation-of-p-o-mppt-based-zeta-converter-fed

GRIN - Your knowledge has value

Since its foundation in 1998, GRIN has specialized in publishing academic texts by students, college teachers and other academics as e-book and printed book. The website www.grin.com is an ideal platform for presenting term papers, final papers, scientific essays, dissertations and specialist books.

Visit us on the internet:

http://www.grin.com/

http://www.facebook.com/grincom

http://www.twitter.com/grin_com

Implementation of P&O MPPT based Zeta Converter fed Three-phase Induction Motor

K.R.Sasitharan,
M.E Scholar; Dept. of EEE,
Government College of Technology,

Dr.K.Ranjith Kumar,
Assistant Professor; Dept. of EEE,
Government College of Technology,

II. BLOCK DIAGRAM

Abstract- This paper proposes use of Zeta Converter as a single stage power conversion concept for adjustable speed drives in photovoltaic applications. Solar panel efficiency is low, To track the maximum power from panel, Perturb & Observe (P&O) algorithms used in Maximum Power Point Tracking (MPPT). Output Torque-Speed characteristics of Induction Machine has been modelled and compared by MATLAB/Simulink software.

Key words: PV Array, P&O method, with MPPT state, Induction Motor (IM).

I. INTRODUCTION

The best way to utilize the electric energy produced by the PV array is to deliver it to the AC mains directly, without using battery storage. The absence of fuel cost, noise, pollution the solar energy source is using for renewable energy among all other sources. The maintenance cost less. However, the PV system has low efficiency due irradiation and temperature. To improve the efficiency of PV system, Maximum Power Point Tracking (MPPT) has been developed such as P&O, Constant Voltage and so on. Previously dc source will be a supply and power to quasi-Z-source inverter with RL load. In this paper solar panel will be an input source to quasi-Z-source inverter with MPPT P&O Techniques with Induction motor have been simulated.

The block diagram of overall PV conversion using MPPT is show in figure 1. The generated voltage (V_{PV}) and current (I_{PV}) from PV array are input for MPPT control. This MPPT control block is calculate the reference voltage. By comparing the reference voltage and PV voltage, the switching pulses (driver signals) generated to switch ON the converter. The converter DC-AC or AC-DC which is depends on the load.

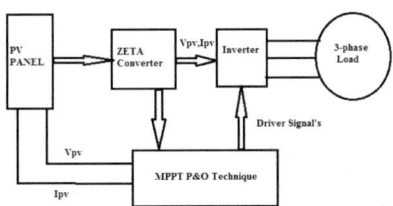

Figure 1: Block diagram of overall PV conversion using MPPT

III.MODELLING OF PV ARRAY

The equivalent circuit of a PV cell is as shown in Figure 2.Where I_{ph} represents the cell photo current, I_0 represents the diode saturation current, I and V are cell output current and cell output voltage respectively. R_p is shunt resistance. R_s are series resistance. They ideal PV module for one diode circuit.

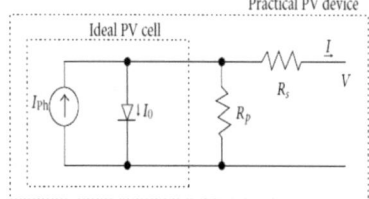

Figure 2: PV cell model

The mathematical model of PV array for single diode circuit can be represented by the following equation [1]:

A. Photo Current (I_{ph}):

I_{ph} depends on the solar irradiation and cell's operating temperature according to the below equation.

$$I_{ph}=[I_{sc}+K_1(T_c-T_{ref})]*H \qquad (1)$$

B. Reverse Saturation Current (I_{rs}):

Reverse saturation current of PV system can be determined by the given equation.

$$I_{rs} = \frac{I_{sc}}{\left[\exp\left(\frac{qV_{OC}}{N_S kAT}\right)-1\right]} \qquad (2)$$

C. Diode Saturation Current (I_0):

Saturation current of PV system varies with the cell temperature can be determined by given equation.

$$I_0 = I_{rs} * \left(\frac{T}{T_r}\right)^3 * \exp\left[\left(\frac{(q*Ego)}{(A*k)}\right)*\left(\left(\frac{1}{T_r}\right)-\left(\frac{1}{T}\right)\right)\right] \qquad (3)$$

D. output current (I):

The equation for output current of the PV system of single diode model presented in Figure 1 is given by,

$$I_{pv} = N_p * I_{ph} - N_p * I_0 \left[\exp\left(\frac{q*(V_{pv}+I_{pv}*R_s R_s)}{N_s*A*k*T}\right)-1\right] - \frac{V_{pv}+(I_{pv}*R_s)}{R_{sh}} \qquad (4)$$

From the above equations,

I_{sc} is cell's short circuit current(A),K is the temperature coefficient(0.0017A/K),T_c is the operating temperature(°C), T_{ref} is the reference temperature(°C), H is solar isolation (kW/m²), q is charge of electron (1.6×10^{-19}C), V_{oc} is open circuit voltage(V), N_s is number of cells connected in series(36), k is Boltzmann constant(1.38×10^{-23} J/K), A is ideal factor(1.6), Ego is band gap energy(1.1eV), N_p is number of parallel connection of cell(1).

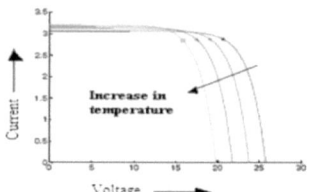

Figure 3: I-V curve of solar cell

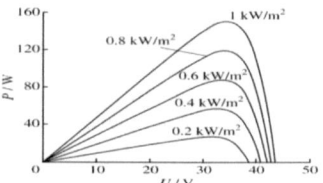

Figure 4: : P-V curve of solar cell

In this paper **Trina solar 240Wp PV module** is taken and the name-plate details are given in Table 1.

Table 1: electrical characteristics data of Trina solar 240Wp PV module

Description	Rating
Rated Power	240Wp
Maximum Power Voltage(V_{mp})	40.5 V
Maximum Power Current (I_{mp})	5.93 A
Open Circuit Voltage (V_{OC})	48.6 V
Short Circuit Current (I_{SC})	6.30A

IV. MPPT ALGORITHM

To improve efficiency of solar panel Maximum Power Point Tracking (MPPT) is used. It includes electronic system to operate the PV modules in a manner that allows the modules to produce all the power. To track the maximum power several techniques are used. The most popular techniques are Perturb & Observe (P&O), Incremental Conductance(IC), Constant Voltage (CV), Open Circuit Voltage, Neural networks and Fuzzy logic. All these methods have their own advantages and disadvantages. The choice of the algorithm depending on the implementation cost, time, and complexity, efficient to track maximum power . In this paper Perturb and Observe (P&O) technique are used with constant and variable duty cycle respectively.

P&O METHOD

It is simplest method of MPPT to implement. In this method output power of solar is checked with the previous output power. If the voltage increasing with power increases then the duty cycle D is increased. For voltage decreasing with power increases duty cycle D is decreased.

The entire process shown as flowchart in Figure 5.

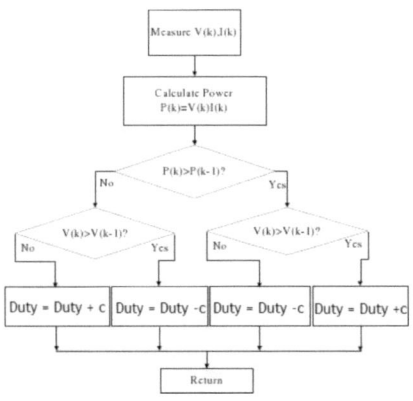

Figure 5: Flow Chart of P&O MPPT

V. INDUTION MOTOR

An electric motor is a device which converts an electrical energy into mechanical energy. This mechanical energy then can be supplied to various types of load. As ac supply is commonly available. The ac motors are classified as single and three phase induction motors, synchronous motors and some special purpose motors. The important advantages of three phase induction motors over other types are self-starting property, no need of starting device, higher power factor, good speed regulation and robust construction. There are two types of rotor constructions which are used for induction motors are,

1. Squirrel cage rotor and

2. Slip ring or Wound rotor.

Here we used in Squirrel cage induction motor and its performance can be analysed through in MATLAB/Simulink.

VI. SIMULATION RESULTS

The simulation of MPPT techniques was carried out and plotted graph for induction motor with stage performance and the simulated circuits.

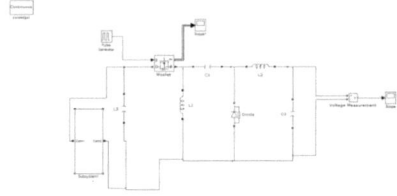

Figure 6: Simulation of Zeta Converter with P&O MPPT technique

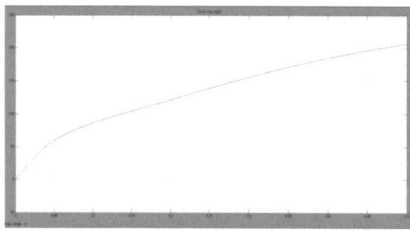

Figure 7: Simulation result of Zeta Converter with P&O MPPT technique output

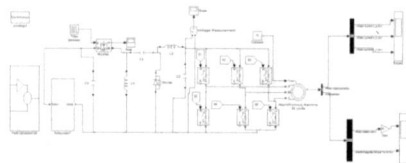

Figure 8: Simulation of Zeta Converter with P&O MPPT method and Induction Motor

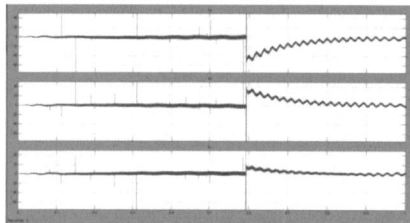

Figure 9: Simulation result of Zeta Converter with P&O MPPT method and Induction Motor stator output

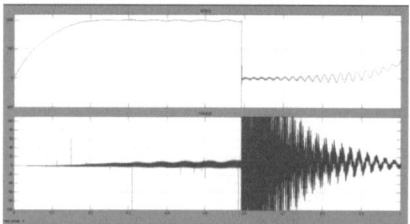

Figure 10: Simulation result of Zeta Converter with P&O MPPT method and Induction Motor mechanical output

VII. CONCLUSION

In this paper, analysed the performance of Zeta Converters in with P&O MPPT technique in Induction Motor and the simulation designs are simulated using MATLAB/Simulink.

REFERENCES

1. F.Baabjerg, A.Consoli, J.A.Ferrira, J.D.Vanwyk. **"The future of electronic power processing and conversion"**, IEEE Transactions on industry Applications,vol.41,no.1,2005,pp.3-8.
2. B.K.Bose, **"The past , present and future of power electronics-guest introduction"**, IEEE on industrial electronics magazine, vol.3,no.2,2009,pp. 7-11.
3. B.Bugar, **"Advanced solar power electronics"**, in proc. Of international symposium on power semiconductor devices and IC's- ISPSD,2010,pp. 3-10.
4. E.Niculescu, **"modeling the pwm zeta converter in discontinuous conduction mode"**, electro technical conference,2008,pp. 651-657.
5. Mathew C.Kessler, **"synchronous inverse SEPIC using the ADP1870/ADP1872 Provides high efficiency for non-inverting buck/boost applications"**AN-1075,pp. 1-7.
6. W.Gu, **"Designing a SEPIC Converter"**, national semi-conductor, application note 1484,june 2007, pp. 1-7.
7. R.C.Viero, **"Dynamic modeling of a sinusoidal inverter based on zeta converter working in DCM for PV arrays"**, annual conference on IEEE industrial electronics socity,2010,pp. 439-444.